Alan,
Merry Christmas
from Grandpa Lutz.

Shadow on the Moon

Shadow on the Moon

A Child's Guide to the Discovery of the ~~Universe~~ Solar System

Char Matejovsky

Robaire Ream

For Anneke, Benjamin, Leighton, Lila, Marlo,
Molly, Sam, Warren and their peers

Library of Congress Cataloging-in-Publication Data

Matejovsky, Char, 1949- author.

Shadow on the moon : a child's guide to the discovery of the solar system / by Char Matejovsky ; illustrated by Robaire Ream.

pages cm

Summary: "Is the earth flat, a square, or a sphere? Geocentric or heliocentric? How do we know? Using rhyme and imaginative illustrations, Shadow on the Moon takes readers on a journey of discovery of the universe from Aristotle and Aristarchus, through the Ptolemy, Copernicus and Galileo, to Kepler and Newton."-- Provided by publisher.

Audience: Ages 8-11.

Audience: Grades 4 to 6.

ISBN 978-1-59815-151-0 (alk. paper)

1. Astronomy--History--Juvenile literature. 2. Astronomers--History--Juvenile literature. I. Ream, Robaire, illustrator. II. Title.

QB46.M38 2014

520.9--dc23

2014025309

Back in the days
before the Common Era had begun,
philosophers reflected
on the planets and the sun.
The sun is hot.
The moon is not.
They all agreed on that.
But does the earth go round the sun?
And is our planet flat?

Well, this planet we call home is round,
so Aristotle said.
The evidence is there to see,
just look up overhead.
The earth's not square.
The earth's not flat.
An eclipse makes it clear.
The shadow on the moon is curved.
That means the earth's a sphere.

Ἀριστοτέλης

ARISTOTLE 384–322 BCE

And his concept of the universe
put all things in their place.
For perfect circles out in space,
he strongly made the case.
The earth sits still. The stars go round
in peace and harmony.
And this they all agreed
is how the universe should be.

SPHERE OF THE PRIME MOVER

FIXED STARS

SATURN

JUPITER

MARS

SUN

VENUS

MERCURY

MOON

EARTH

περὶ
οὐρανοῦ

350 BCE

Ἀρίσταρχος

ARISTARCHUS

310-230 BCE

JUPITER

VENUS

MERCURY

SUN

EARTH

MOON

SATURN

MARS

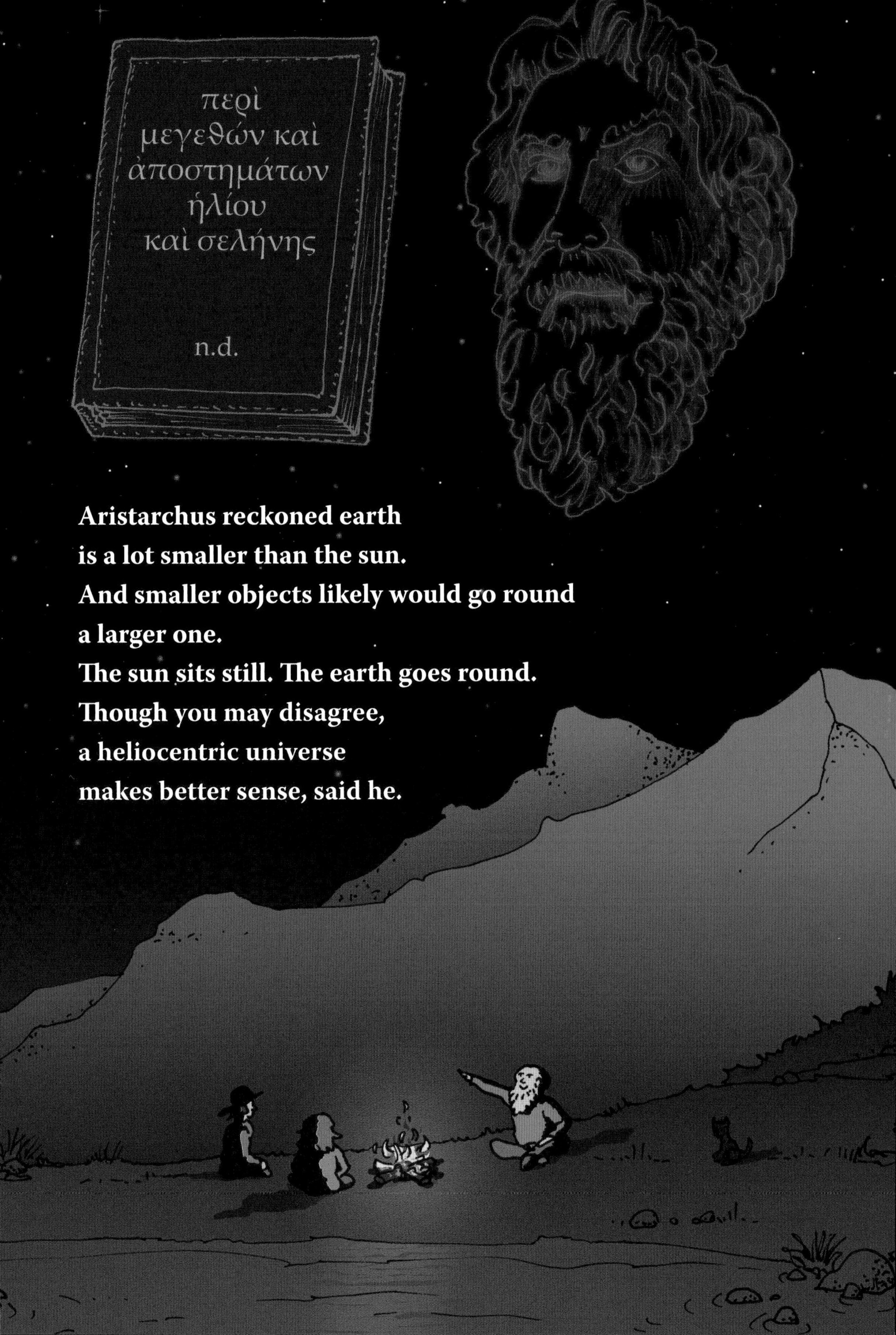

Aristarchus reckoned earth
is a lot smaller than the sun.
And smaller objects likely would go round
a larger one.
The sun sits still. The earth goes round.
Though you may disagree,
a heliocentric universe
makes better sense, said he.

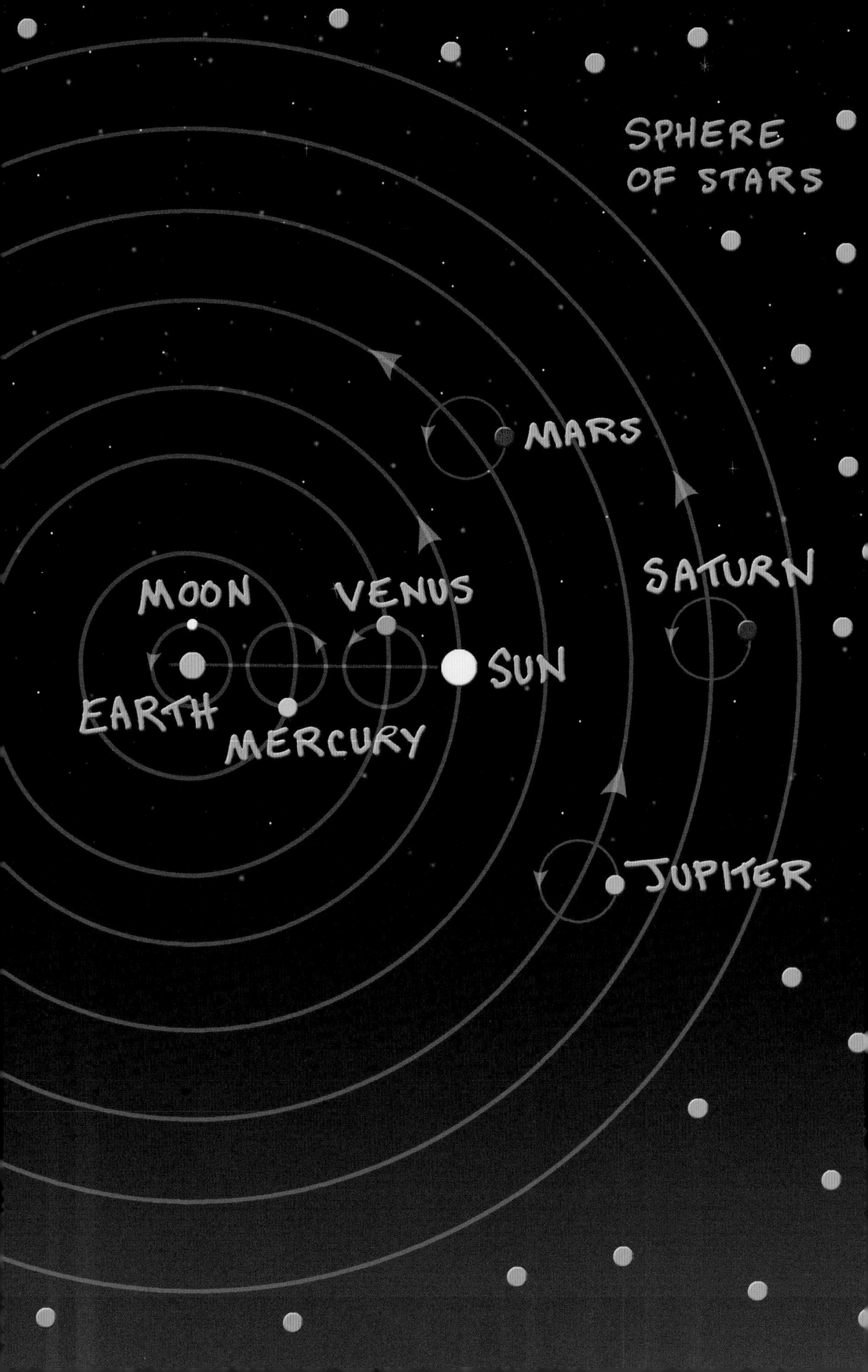

SPHERE OF STARS

MARS

SATURN

MOON

VENUS

SUN

EARTH

MERCURY

JUPITER

An astronomer named Ptolemy
then stepped into the fray.
His theory of the universe
for centuries won the day.
Earth centered still, it filled the bill,
though simple it was not.
With circles within circles
he found answers people sought.

Πτολεμαῖος

PTOLEMY ca 87–150 CE

But the geocentric universe
that Ptolemy devised
could not explain some data
and would need to be revised.
A thousand years. Four hundred more.
It really took that long
before a mathematician charged
that Ptolemy was wrong.

384 BCE
Aristotle was born

Common Era begins

310 BCE
Aristarchus was born

ca 87 CE
Ptolemy was born

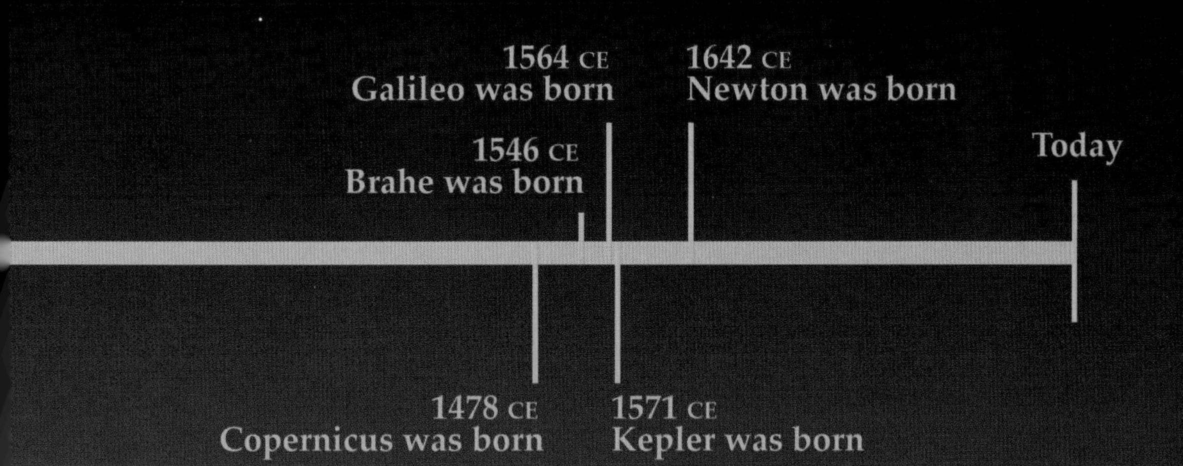
1564 CE
Galileo was born
1642 CE
Newton was born
1546 CE
Brahe was born
Today
1478 CE
Copernicus was born
1571 CE
Kepler was born

Nicolaus Koppernigk

Now one problem that Copernicus
would tackle was the way
that planets could go forward
then go back another day.
From west to east. Then east to west.
They knew that couldn't be.
It's simple. Earth is moving
and we're passing them, said he.

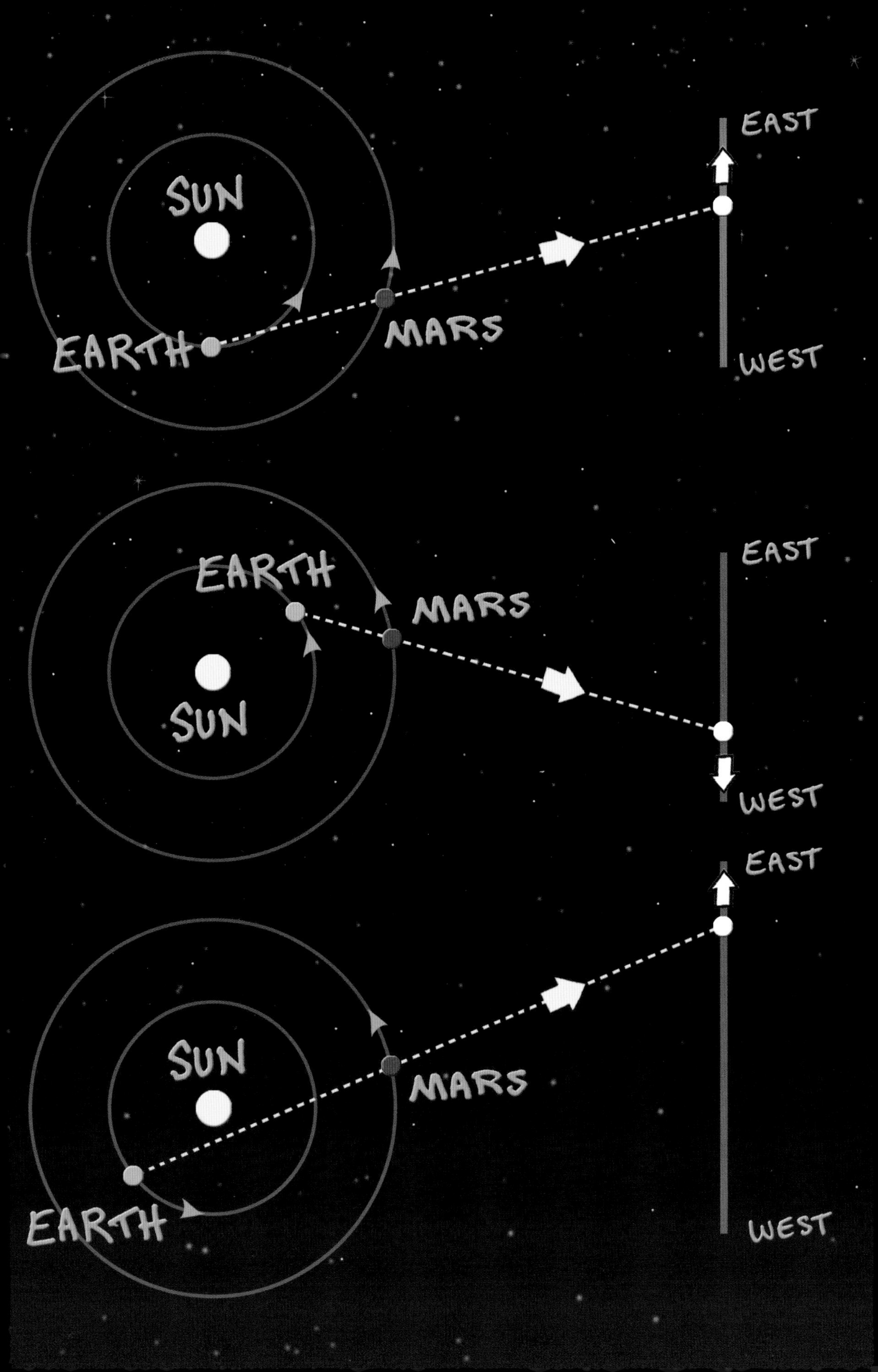

SUN
EARTH
MARS
EAST
WEST
EARTH
MARS
SUN
EAST
WEST
SUN
MARS
EARTH
EAST
WEST

SATURN
JUPITER
MARS
MOON
EARTH
VENUS
MERCURY
SUN
FIXED STARS

And the earth is not the center, no,
when all is said and done.
It's just another planet,
the third outward from the sun.
The center is the sun, he wrote,
the planets circle it.
Not so, replied the critics,
you've no evidence to fit.

Galileo Galilei

1564–1642 CE

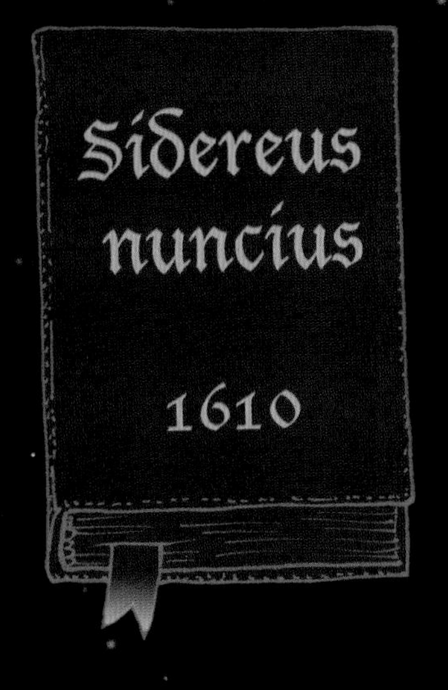

Galileo got a telescope
and looked up at the skies.
He trained his sights on Venus,
and he came up with the prize.
In crescent here. In full phase there.
His spyglass let him see
that Venus travels round the sun,
as clear as clear as can be.

DEL SIG. GALILEI. 217

ducete ſino à dannar con lunghi diſcorſi chi prende il termine vſitatiſſimo d'infinito per grandiſſimo. Quando noi abbiamo detto, che il Teleſcopio ſpoglia le Stelle di quello irraggiamento, abbiamo voluto dire, ch'egli opera intorno à loro in modo, che ci fà vedere i lor corpi terminati, e figurati, come ſe fuſſero nudi, e ſenza quello oſtacolo, che all'occhio ſemplice aſconde la lor figura. E egli vero Sig. Sarſi, che Saturno, Gioue, Venere, e Marte all'occhio libero non moſtrano trà di loro vna minima differenza di figura, e non molto di grandezza ſeco medeſimi in diuerſi tempi? e che coll' occhiale ſi veggono Saturno, come appare nella preſente figura, e Gioue, e Marte, in quel modo ſempre; e Venere in tutte queſte forme diuerſe? e quel, ch'è più merauiglioſo con ſimile diuerſità di grandezza? ſi che cornicolata moſtra il ſuodiſco 40. volte maggiore, che rotonda, e Marte 60.

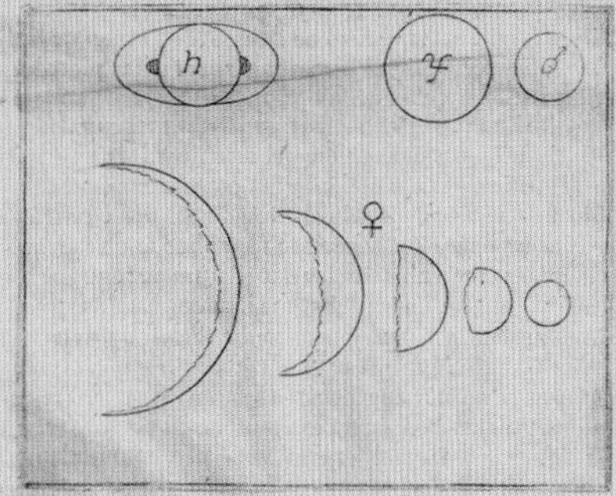

volte, quando è perigeo, che quando è apogeo, ancorche all'occhio libero non ſi moſtri più che 4. ò 5.? Biſogna, che riſpondiate di ſi, perche queſte ſon coſe ſenſate, ed eterne, ſi che non ſi può ſperare di poter per via di ſillogiſmi dare ad

E e inten-

Johannes Kepler

1571-1630 CE

And along came Kepler with the data
Brahe had amassed.
The model of the universe
once more would be recast.
I've studied Mars at length, he said,
and I've discovered this.
Its orbit is not circular,
it's really an ellipse.

$$\frac{T_1^2}{T_2^2} = \frac{R_1^3}{R_2^3}$$

THIRD LAW OF PLANETARY MOTION

SATURN
JUPITER
VENUS
MARS
EARTH

SUN
MERCURY

Isaac Newton

1642–1727 CE

Now, if planets move about in space,
why don't they fly away?
mused Newton as he watched an apple fall
one summer day.
It's gravity, of course, he cried!
It even works in space.
The bigger mass, the bigger force.
The sun keeps us in place.

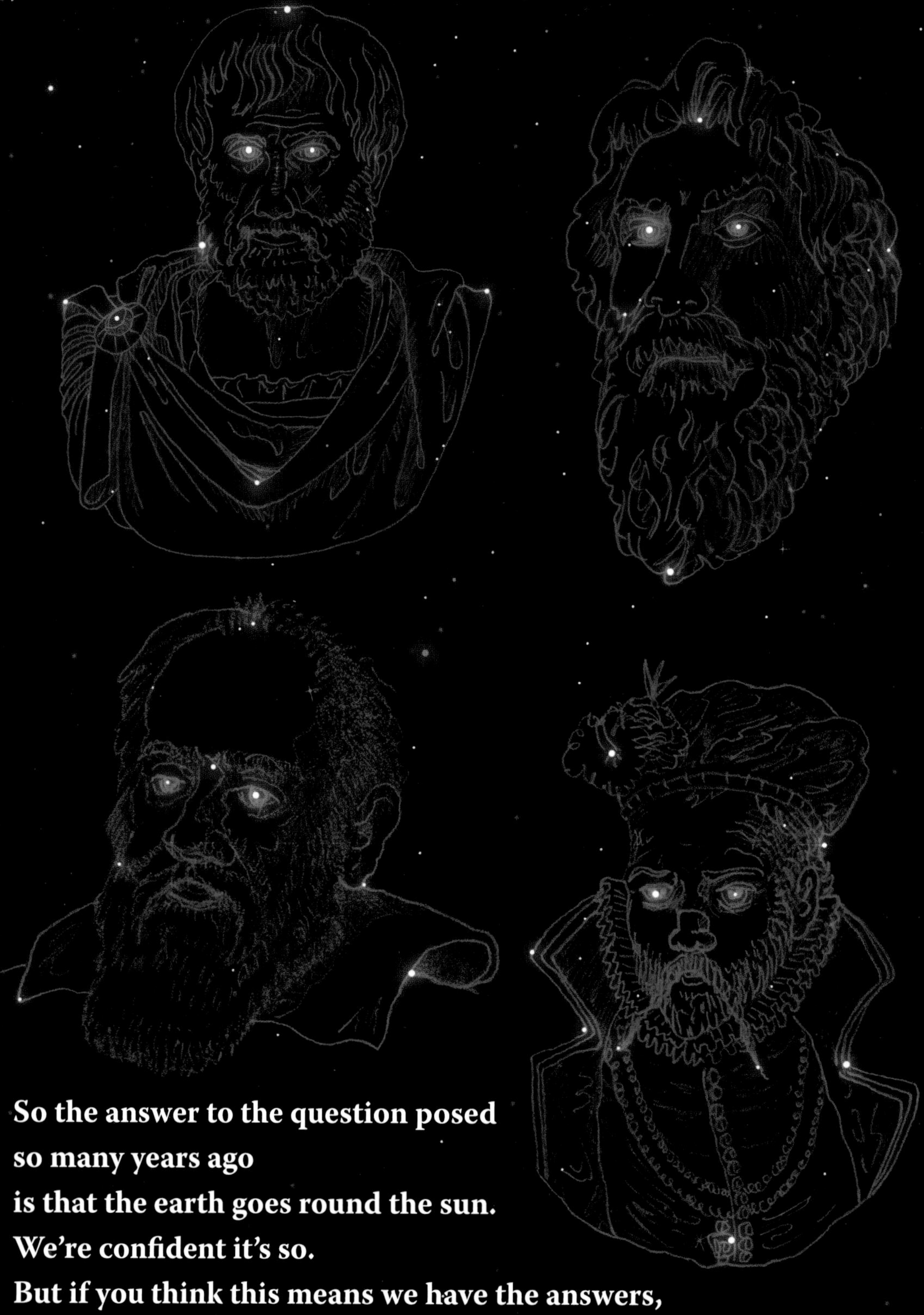

So the answer to the question posed
so many years ago
is that the earth goes round the sun.
We're confident it's so.
But if you think this means we have the answers,
think again.
We surely have more questions now
than they had way back then.

For what seemed to be a universe
up till the recent past
is just one solar system
in a universe so vast
we cannot comprehend its scope,
we do not know its size.
There's still a lot to ponder
when we look up at the skies.

THE SOLAR SYSTEM

A α alpha
B β beta
Γ γ gamma
Δ δ delta
E ε epsilon
Z ζ zeta
H η eta
Θ θ theta
I ι iota
K κ kappa
Λ λ lamda
M μ mu
N ν nu
Ξ ξ ksi
O ο omicron
Π π pi
P ρ rho
Σ σς sigma
T τ tau
Υ υ upsilon
Φ φ phi
X χ chi
Ψ ψ psi
Ω ω omega

Names

Ἀρίσταρχος = Aristarchus
Ah-**ree**-star-kos Air-i-**star**-kus

Ἀριστοτέλης = Aristotle
Ah-ree-sto-**teh**-lays Air-i-**stot**'l

Copernicus
Koh-**per**-ni-kus

Πτολεμαῖος = Ptolemy
Ptoh-leh-**may**-os **Toh**-leh-mee

Titles

μαθηματικὴ σύνταξις
mah-thay-mah-tee-**kay** **soon**-ta-ksis

περὶ μεγεθῶν καὶ ἀποστημάτων ἡλίου καὶ σελήνης
peh-**ree** meh-geh-**thohn** kai ah-poh-stay-**mah**-tohn hay-**lee**-oo kai seh-**lay**-nays

περὶ οὐρανοῦ
peh-**ree** oo-rah-**noo**

Terms

heliocentric = sun centered
geocentric = earth centered

Notes

Aristarchus reckoned that the diameter of the sun was about seven times that of the earth. In reality it is 109 times.

The illusion that planets are traveling backwards is called retrograde motion. Ptolemy's geocentric model had planets move in small circular orbits, called epicycles, so that they could go backwards. Copernicus argued for a simpler, heliocentric model in which planets only appear to go backwards because earth is passing them.

Danish astronomer Tycho Brahe made the precise astronomical observations later used by Kepler as the basis for his laws of planetary motion.